SECURE

insights from the people who keep information safe

Edited by Mary Lou Heastings

EA
EXECUTIVE ALLIANCE

Executive Alliance Publishing House

EXECUTIVE ALLIANCE PUBLISHING HOUSE

A Division of Executive Alliance, Inc.

3605 Sandy Plains Road

Suite 240-429

Marietta, Georgia 30066

Copyright © 2010 by EXECUTIVE ALLIANCE PUBLISHING HOUSE

All rights reserved,
Including the right of reproduction in whole or in part in any form.

EXECUTIVE ALLIANCE PUBLISHING HOUSE
are trademarks of Executive Alliance, Inc.

Printed in the United States of America

Book Design, Cover Design & Illustration by Scott Cathey
scottcathey.com

ISBN 978-0-9769173-3-5

Table of Contents

Acknowledgments - page vi

Foreword - page vii

Leadership Competencies Needed for the Future
Deborah Snyder – NYS Office of Temporary and Disability Assistance (OTDA)
page 1

Developing a Business Continuity Plan in a Mid-Sized Financial Services Company
Howard Israel – Fidessa corp.
page 11

There's No "Point" to Security Anymore
Dave DeWalt – McAfee
Security Leaders Summit Northeast 2010 – Bronze Sponsor
page 19

Discussions with Senior Management on Information Security
Roseann Larson – Estée Lauder
page 25

**Application Delivery Networks:
The New Imperative for IT Visibility, Acceleration and Security**
Blue Coat Systems
Security Leaders Summit Northeast 2010 – Bronze Sponsor
page 31

Trends – Insights into Preparing for the Future
John Walp – M&T Bank
page 45

Why Enterprises Need a Flexible Approach to Data Security
Sean Glynn – CREDANT Technologies
Security Leaders Summit Northeast 2010 – Bronze Sponsor
page 53

Acceleration of Change in Information Security
Roy L. Post – AXA Equitable
page 59

Strategies for Building a Successful Roadmap
Todd Colvin – Paychex, Inc.
page 69

Secure by Design: A Foundation to Software Innovation
Al Zollar – IBM Tivoli
Security Leaders Summit Northeast 2010 – Platinum Sponsor
page 79

Acknowledgments

I would like to thank each of the authors who contributed to this book by sharing their insights into different topics. I have enjoyed working with each of you in putting the content together.

Thank you to Scott Cathey, our graphic designer, who created our cover design and pulled all the interior content together to help make this book happen.

And to the Executive Alliance team, a special thank you for your hard work and dedication to getting all things right.

Mary Lou Heastings
CEO, Executive Alliance

....
"By contributing to this book, it is not an endorsement of a company's product or solution contained within the book."

Foreword

The professionals represented in this book bring years of experience, industry knowledge, and success to their organizations. Focused on delivering higher levels of value to their organizations and their customers, both practitioners and solution providers share their insights into where they have seen success and where they believe future attention should be placed. Whether it is having the right discussions with senior management, building a successful road map or business continuity plan, or evaluating how you design and implement your security programs, this book touches on many of the topics that are of concern and interest today to information security professionals.

SECURE

Leadership Competencies Needed for the Future

Deborah Snyder, CISSP, GIAC GSLC, PMP

Chief Information Security Officer – NYS Office of Temporary and Disability Assistance (OTDA), Division of Legal Affairs
Headquarters: Albany, NY

The New York State Office of Temporary and Disability Assistance (OTDA) is responsible for supervising a wide range of the State's most important programs that provide assistance and support to eligible families and individuals. These programs serve to enhance the economic security of low-income working families, assist work-capable public assistance recipients in achieving entry into the workforce, assist individuals with priority needs other than work-readiness in accessing appropriate benefits and services, enhance child well-being and reduce child poverty

....

Successfully leading an organization's information security program requires a strong combination of leadership and management skills, knowledge across both non-technical and technical areas, and the ability to apply these competencies to the organization's goals and initiatives.

While there are thousands of books, papers, articles and blogs on leadership, and the traditional skills associated with a good leader, how do such qualities relate to successful and effective leadership in the field of information security? Which specific leadership abilities and skills contribute to success in this critical and dynamic environment, and what competencies[1] will information security executives and professionals need to be relevant and effective going forward?

The following leadership traits and proficiencies are essential and will help you effectively lead your organization's information security program, lead others, and lead yourself:

BUSINESS ACUMEN AND ALIGNMENT

A clear understanding and sense of your organization's mission, business and direction, and the ability to make sound judgments and decisions related to it, are some of the most critical competencies for information security leaders.

Information is the lifeblood of organizations – a vital asset that must be properly managed and safeguarded. Organizations are highly dependent on data and the mechanisms through which information is gathered, stored, distributed and utilized to support business functions. The goal of information security assurance is to protect information assets against risk and maintain their value to the organization.

Alignment with the organization's business goals and objectives is key to getting security right - getting the right information to the right people at the right time, and ensuring synergy across policies, controls, training and technology solutions. Maintaining the appropriate security posture serves and enables business. Ensure your organization's business needs are the primary driving force in considering security-related decisions and investments.

A firm grasp of resource management, budget, finance and procurement processes are also important, since they are the foundational to business, and help you prepare and justify resource requests.

....

[1] *Competencies, as generally defined, are behaviors, knowledge, skills and proficiencies that directly and positively influence the success of a leader and their organization. They can be defined, learned, measured and enhanced over time.*

STRATEGIC PLANNING AND MANAGEMENT

Leaders think strategically, prioritize and focus fiercely on the things that matter most, and avoid distractions. They also have the tactical expertise to use available resources effectively to get results in key areas. Ability to see the big picture and a clear vision of the desired end-state is essential.

Lead with purpose. Prioritize information assets based on criticality to the organization, sensitivity of the data, and related risk and control factors. Devise a comprehensive, well-aligned and meaningful enterprise information security strategic plan. It serves as a roadmap and framework that lays out goals, objectives and actionable steps to achieve them. It sets forth a portfolio of initiatives that support the organization's ability to safeguard information assets, manage risk, assure compliance, and prevent and address incidents. Draw upon project management expertise to efficiently and successfully complete planned initiatives.

Assure your organization's information security strategic plan also considers global cyber security concerns. Stay current on evolving security threats. Keep a strategic "eye-on-the-horizon" view and hone your ability to forecast potential impact, to help your organization stay ahead of the risk curve.

COMMUNICATION SKILLS

Leaders must have exceptional communications skills (verbal, written, presentation), and the ability to clearly, concisely and effectively convey information in both technical and non-technical terms. Strong communications skills are key to obtaining support and commitment at all levels – from your executives, across business units and end-users, on what needs to be done and why it is important.

Create scenarios that allow people at every level to understand security-related concerns, potential impact on business functions, and available strategies. They help you "tell the story," get issues and ideas across, and sell the value proposition of security investments. Building a security-aware work force goes well beyond ensuing employees adhere to policies. You must create an organizational culture of awareness and individual responsibility for information security.

RISK MANAGEMENT

Leaders recognize potential threats to their organization's business functions and assets, and understand their organization's appetite or tolerance for risk. Business objectives often present a tricky balance between risk and reward. Organizations want to capitalize on opportunities to do business in new ways, enhance capabilities and create efficiencies. Such changes can pose increased risk to critical information assets.

Lead your organization's management of information security risk. Work with organizational partners to identify, analyze and understand risks. Prioritize risks based on potential probability and impact on the business. Provide guidance as to available control options. Effectively implement measures that best fit your organization's requirements, resources and environment. Leverage industry standards to reduce cost and effort. Be proactive in helping your organization strike the appropriate balance between doing business and safeguarding information, so the business can move forward as desired.

Caveat: be realistic. It is not possible to eliminate all information security-related risks. Ensure your executives fully understand and sign off on the residual risk related to business processes and initiatives.

COMPLIANCE ASSURANCE

Most organizations are held to a myriad of laws, regulations and requirements related to confidentiality, privacy and information security. Leaders understand that compliance mandates are critical business drivers that must be understood and addressed to avoid potential penalties and liabilities.

Ensure you have a keen sense of the statutory, regulatory and policy requirements that affect your organization. Know how they are orchestrated through policies, standards, controls and related auditing protocols. Use standards to help assure consistency and acceptance. Know how compliance is monitored and enforced. Document well, and be prepared to respond to routine audits and findings.

BUILDING NETWORKS & PARTNERSHIPS

Leaders excel at building meaningful interpersonal relationships and partnerships. They leverage these alliances to capitalize on collective knowledge, collaborate and learn from others, and identify best practices.

Recognize the importance and relevance of building professional relationships. Develop networks and partnerships, and work to obtain cooperation and commitment from others. Empower others by sharing power and responsibility, and establishing and maintaining rapport with key players.

COLLABORATION AND NEGOTIATION SKILLS

To encourage collaborative thinking, leaders create an environment that supports creativity, diverse perspectives and approaches, and promotes honest input, fairness and compassion.

Security decisions cannot be made in a vacuum. You must be able to build bridges and close gaps between business units, legal, contracts and IT, and effect collaboration both within and beyond your organization. Convey openness, and a willingness to listen and look beyond the status quo. Be receptive to exploring alternatives and new ideas. Facilitate active discussion, seek out and value the collective knowledge and opinions of others. Demonstrate sensitivity to diversity in culture, race, gender, background, experience and other individual differences. Guide others to see the value of diversity, and building a positive and healthy working environment.

INFLUENCE, CONSENSUS BUILDING AND CONFLICT RESOLUTION

Leaders can effectively guide collaborative thinking to consensus and resolve conflicts. They have the ability to get others to listen, to convince people of the importance of concerns and the value of changes and investments.

You must connect with, inspire and motivate people to rally around complex security issues and initiatives, to achieve desired outcomes. Energizing and convincing others to embrace change and take on work can be particularly difficult in tough times, when resources are lean and everyone's plates are already too full. Know when to press on, how to push back, and choose your battles carefully. Ensure conflicts are resolved in a professional and positive manner, so that all feel heard and gain something from the experience. A security-aware organization involves everyone pulling together around common concerns, and helps best serve your organization.

ACCOUNTABILITY, INTEGRITY AND AUTHENTICITY

Successful leaders demonstrate firm belief in their abilities and ideas, and take ownership for their areas of responsibility and accountability for effectively serving their organization. Their actions and professional ethics reflect solid core values. They are organized, self-motivated and results-oriented, and hold themselves accountable for their own performance. They emphasize personal character development and use their position and personal power appropriately.

To lead, you must possess self-awareness and self-confidence. Have faith in yourself. Be approachable, honest and open. Take a pulse check often to assure you are holding yourself accountable. Ensure your values align with organizational values and reconcile disparities to operate most effectively.

As a leader, you also extend your personal self-awareness to those around you. Be sensitive to the impact of your behaviors on others. Build credibility and trust by following through on your commitments.

Exercise humility, recognizing that you cannot do the job alone. Demonstrate interpersonal understanding, and a deep respect for the value of others. Help build team-based alignment and awareness, and common expectations as to performance and results.

GROWTH POTENTIAL

Leaders exhibit a willingness to continuously learn through multiple channels including research, formal education, skills training and networking. They recognize personal strengths and weaknesses, work to improve weak points, and involve appropriate subject matter experts to ensure these areas do not affect overall effectiveness.

To lead yourself, you must be willing to step outside your given comfort zone. Roll up your sleeves and work to enrich your knowledge in areas, business processes and technologies that you are not fully versed in. Seek out personal mentors, particularly in areas where you could benefit from additional experience and expertise.

To lead others, you must have strong team-building, coaching and mentoring skills. Assist individuals in developing themselves, drawing on your experience and knowledge. Help team members achieve their professional potential. Provide objective feedback as to strengths and areas for improvement. Identify and help others develop the necessary competencies to accomplish future goals and support your organization.

FLEXIBILITY & AGILITY

Leaders view and expect change as a constant. Emerging threats, new compliance mandates, changes in the ways in which your organization and the world in general conducts business, technologies and public expectations, can all create security-related risk and impact.

Be resourceful and have the capacity to respond and adapt quickly to new and changing threats, requirements, technologies and issues.

Know what is going on, proactively embrace change and lead the transition - in yourself, in those around you, in your organization and in the industry.

PERSISTENCE AND RESILIENCY

Leaders have a deep sense of conviction and commitment to what they do, and a driving sense of urgency and passion for its importance to the organization.

You must be able to face and overcome resistance and adversity, whether it is related to change, difficult people, challenges or tough times. Be consistent. Keep going despite obstacles, have the courage to see beyond and the conviction to push through. Managing incidents and functioning well in a crisis is the ultimate test of a leader. Remain calm, act professionally, think and communicate clearly. Demonstrate fortitude, grace under fire and the tenacity to steer your organization through times of tremendous change, chaos and disruption.

OPTIMISM AND ENTHUSIASM

Leaders are a visibly positive force for those around them. To lead, you must maintain a positive outlook, expend your best effort, refuse to act the victim of circumstances, find creative ways to manage around impediments and get things done.

You must also possess incredible, infectious, unstoppable, firing-on-all-cylinders energy! Tirelessly go the extra mile, set the example and demonstrate commitment to your organization.

Recognize that you have got your work cut out for you. Accept that you will face trials and uphill battles, and have days when it seems like nothing you do or say makes a difference. Keep the end in mind, keep smiling and stay positive anyway!

In the end, the key to leadership success in information security is to work hard, communicate effectively, stay focused, be true to yourself, treat others well, stay positive and keep going despite challenges. If you practice and demonstrate these leadership traits and behaviors, you will serve your organization well, and enjoy success in one of the most dynamic, challenging and rewarding of professions! Godspeed!

About Deborah Snyder

Ms. Snyder has over twenty-five years of experience in human services programs and information systems, meeting business needs through the innovative use of technology, mission-critical redesign and modernization projects, and strategic policy planning and implementation.

In her current role, she directs the agency's Information Security Office and oversees all aspects of its comprehensive Information Security Assurance Program - a portfolio of initiatives aimed at protecting State and agency information assets, identifying and mitigating risk, assuring federal and state compliance, developing security policies/ procedures, improving information security services and awareness, and enhancing the ability to manage risk, and prevent, detect and recover from incidents.

Ms. Snyder holds a B.A. from the State University of New York, and has completed postgraduate work in information security assurance, incident management, evidence handling and digital forensics. She holds several industry certifications including Certified Information Systems Security Professional (CISSP), SANS Global Information Assurance Certification in Security Leadership (GSLC), and Project Management Professional (PMP).

Ms. Snyder serves as Co-Chair of the NYS Forum Security Work Group, and is a member of the Project Management Institute, InfraGard National Members Alliance (INMA) representing over 26,000 FBI-vetted InfraGard Subject Matter Experts, Information Systems Security Association (ISSA), Computer Security Institute (CSI), Institute of Internal Auditors (IIA), NYS Computer Forensics Workgroup and NYS Digital and Multimedia Evidence Technical Working Group. She is also a PMP Prep instructor, and frequently requested speaker at prominent government and industry forums such as the Government Technology Conference (GTC), NYS Cyber Security Conference, NYS CIO Counsel, CIO Academy, and the Multi-State Information Sharing and Analysis Center (MS-ISAC) National Webcast Initiative on information security and risk management topics critical to executive-level, senior business and IT professionals.

SECURE

Developing a Business Continuity Plan in a Mid-Sized Financial Services Company

Howard Israel

Corporate Security Officer – Fidessa corp.
Headquarters: New York, NY

Fidessa group is the leading supplier of multi-asset trading, portfolio analysis, decision support, compliance, market data and connectivity solutions for firms involved in trading the world's financial markets. Fidessa's products and services are built on the simple vision of making it easier to buy, sell and own financial assets of all types on a global basis.

....

THE BUSINESS PROBLEM

Developing a Business Continuity Plan (BCP) for a mid-sized technology company in the financial services industry is challenging primarily because of the extremely time critical nature of the processes involved - business continuity needs to kick in instantaneously if there is a system fail over. Any down time could result in millions of dollars worth of losses and there are also regulatory requirements that must be kept in mind.

This chapter will discuss at a high level how Fidessa implemented its BCP, the challenges it faced and how they were addressed.

Fidessa provides both SaaS and Enterprise solutions to its clients. Smaller customers tend to choose the SaaS solution because of its light footprint and larger clients prefer the Enterprise model which has a large onsite presence and numerous full-time staff. Both require comprehensive and well conceived business continuity plans.

CHALLENGE

Provide high uptime in a demanding industry and satisfy customer BCP and regulatory requirements.

Architecture: Two main data centers running the hosted production systems in a "hot-hot"[1] configuration with dual connectivity. The Fidessa system is designed for automatic failover to the secondary "hot" server. Support systems will failover with minimal administrator effort. An additional office location contains several server rooms that house the office IT and communications systems.

STARTING POINT

A comprehensive BCP requires more than just failover technology. A crucial part of business continuity planning is the plan itself. At Fidessa, the technology infrastructure for the production hosted systems was already designed for BCP. From a technology perspective, some additional build-out of the office infrastructure was required, but the biggest hole was lack of documentation around the BCP.

Thorough documentation should describe how a potential BCP event should be managed, how and when communication should take place and what each department should do to meet their unique business commitments. The goal should be to enable the staff – from senior management to junior employees - to manage and implement the BCP.

---

[1] *This refers to the fact that both data centers are always live however the primary and secondary systems are split between each data center. The major advantage of this type of configuration is that if one of the data centers was completely offline, only half of the systems would need to failover.*

Documentation is futile unless the information is read, understood and digested. This means staff training is necessary at all levels.

The curriculum should reach the following audiences:
- *The highest levels of the senior management who will need to understand the issues in managing an event*
- *The mid/junior staff who will need to practice the technical aspects of failovers*
- *All staffers who will need to understand emergency procedures in terms of work location and their role during an event.*

There are numerous industry initiatives that firms can get involved in to test BCPs. For example, Fidessa has long been a regular participant in the financial industry's "street-wide" BCP tests run by SIFMA[2]. Beyond this measure, additional testing on a regular basis is needed for two reasons: 1) to test the failover capabilities designed within the infrastructure and 2) to have confidence in the staff to diagnose business critical problems and react accordingly.

GAP ANALYSIS

The first step in producing and implementing any BCP is to perform a gap analysis to determine what BCP procedures exist, what needs to be built and the level of testing required. Senior management must buy in to this process as they will be required to own certain areas of the plan.

Senior staff members responsible for each department often have a plan for business continuity in mind but all too often it has not been documented or checked against other departments. In the worst case scenarios the plan hasn't even been communicated to staff or tested in advance. Each piece of the plan must be created, evaluated for

---

[2]*Additional SIFMA "street-wide" BCP testing information can be found at: sifma.org*

potential gaps and fleshed out before an event takes place. Often there are company-wide elements in a BCP that affect every department but aren't assigned to any one division in particular. This overlap and the lack of bottom line responsibility can expose gaps in the plan that need to be defined, such as:

- *A Management Plan – to address how management should function in the event that not all senior managers are available during an event.*

- *A Communications Plan – to address information sharing policies when primary communications are not available or an emergency happens outside of regular business hours.*

- *An Employee Emergency Response Plan – to address employee-specific concerns including where to meet, how to communicate with their manager, what their responsibilities are or how to access company or department-wide conference bridges.*

The points above all concentrate on specific, tactical information: names, telephone numbers, conference bridge info, addresses, locations, driving directions, etc. Once these core elements have been identified, individual departmental plans can be created by the department heads and their staff.

DEPARTMENTAL PLANS

Departmental plans often take the most effort to create and more importantly require mutual cooperation between various team members. A sample BCP should be designed and rolled out to different departments. The creators of the plan must be sure that it is clear and can be easily tailored to capture the specific information that is unique to each department. By providing a basic template with sample information department heads can focus on adding the information their divisions require. As plans are drafted, departmental inter-dependencies and gaps in current practices can become apparent and will need to be coordinated. As department heads document their IT infrastructure and the other services they deem critical there can be missing information around how to address potential outages. Each discrepancy needs to be tracked as a separate action item for follow-up. As such, the development of these department plans can become an evolution. The departments with deep sets of existing processes must be particularly diligent in adding crucial details. Additionally, each department should document the critical external suppliers that wouldn't necessarily be apparent to someone stepping in to manage an event or failover situation. When all of the department plans have been drawn up, they need to be sanity checked against one another to identify further gaps and inconsistencies.

THE BCP PACKAGE

The BCP is actually a package of multiple short tactical documents which consists of the following elements:

- *A Management Plan*
- *Various Departmental BCP's*
- *A Employee Emergency Response Plan (this is the only document that should be available to all employees)*

BCP BRIEFINGS

This should be a two-phased approach. First, the department heads should be briefed on their responsibilities in terms of managing an event (this should come from the content of the Management Plan and the Communications Plan)[3]. Second, they will need to be instructed on how to brief their staff. The staff briefings will address the common company-wide information that each employee needs and the

....

[3]*The communications plan is documented within the management plan and the employee emergency response plan and contains information that is appropriate to each different audience.*

department-specific plan at the same time. Having the department head deliver the general company information helps ensure that they themselves are familiar with this information.

BCP TESTING

BCP testing needs to be focused on the most critical areas of the plan. This serves two purposes: to ensure the technology and processes work correctly and to ensure that staff members have sufficient knowledge to execute the failover process.

The test plan generally consists of a simulated outage, diagnosis and creation of an action plan. Following the necessary approvals of the plan elements, senior managers should execute the action plan against a simulated failover by testing and implementing every documented procedure.

Each test is followed up with a post-mortem meeting where issues that have been uncovered can be discussed and acted upon. The actions fall into three categories:

- *Technical*
- *Process (documentation)*
- *Training*

BCP EVOLUTION

When the BCP is completed, the next step is to start the maintenance process: work on the BCP should be an ongoing process. The plans will naturally become obsolete over time. As technology, processes and staff change so the BCP will need to be updated. One mechanism to address this would be to keep track of the needed changes, in a Wiki or something more formal, for example, so that they can be addressed when required.

A FINAL NOTE

The above case study describes at a very high level the BCP creation process. Naturally, as this process is likely to involve a significant number of senior staff over a several month period, some details have been omitted. Nevertheless, it is hoped that this provides a sufficiently detailed roadmap to allow other organizations to create their own BCP plans.

....

About Howard Israel

Howard is currently the Corporate Security Officer of Fidessa corp. He has been in the IT Security field since 1982 having worked at the US DoD National Security Agency, AT&T Bell Labs, AT&T UNIX System Labs (USL), AT&T Taiwan, AT&T WorldNet, Lucent, Avaya and as an independent contractor to Johnson & Johnson. He has published various technical papers and several government reports and standards on security. His current responsibilities include IT Security, Physical Security, and Business Continuity Planning for Fidessa corp.

SECURE

There's No "Point" to Security Anymore

Dave DeWalt

President and Chief Executive Officer – McAfee
Headquarters: Santa Clara, CA
Security Leaders Summit Northeast 2010 – Bronze Sponsor

McAfee is the world's largest dedicated security technology company. We relentlessly tackle the world's toughest security challenges. McAfee's comprehensive solutions enable businesses and the public sector to achieve security optimization and prove compliance and we help consumers secure their digital lives with solutions that auto-update and are easy to install and use.

....

The era of providing complete protection by installing multiple best-of-breed security products has passed. Today's world of sophisticated malware, targeted threats and multi-stage attacks requires security that is smart, cross-linked and interoperable. Security must extend well beyond the traditional disparate defenses that form a broken safety net made up of point tools. Let's look at the past year and discuss two cyberattacks that gained a lot of attention and made headlines around the world. Why were some of the targets in the attack protected, while others weren't? The answer lies in what I call "Global Threat Intelligence."

On July 4th, 2009, a botnet was used in what's now known as one of the most serious distributed denial-of-service attacks in recent history. It was a day when many Americans were BBQ-ing and many security operations centers were thinly staffed. The attackers designed a particularly nasty piece of malware that created a botnet of about 200,000 machines. Not only could the malware render the hardware of a PC inoperable by deleting key files and overwriting the master boot record, but it had automatic update functionality as well.

Several U.S. government websites came under attack from the botnet. Most notably the FTC, FAA and Treasury were hit, but the attack was not limited to these websites and would soon spread to both government and commercial websites in South Korea and the United States. We know that some of the target websites went down during the periods of attack. Some people associated the attacks with North Korea and called them an act of cyberwar.

So who was protected and why? One of the answers lies in understanding the reputation of the IP addresses in the botnet. The systems involved in these damaging attacks had provided clues that they were up to no good. Before the systems were part of those damaging attacks, something else happened. The operators of the botnet used it for various nefarious activities. One of those was sending out spam. McAfee captured spam from the offending IP addresses and the reputation of these addresses was downgraded within our Global Threat Intelligence system.

What's unique about this? Well, in this real-world attack scenario the enterprise firewall protected customers based on threat intelligence that was gathered earlier by the email security appliance. Two entirely different protection vectors shared threat intelligence through the cloud, and as a result a malicious attack was thwarted.

That's something that is only possible with a correlated defense where different layers of defense talk to each other, share intelligence and provide predictive protection as a result.

More recently, Operation Aurora was an attack designed to steal high-value intellectual property from Google and dozens of other companies. The world learned about Operation Aurora when Google published a blog saying it had been the subject of a sophisticated cyberattack and that various other companies had also been targeted.

The depth and scope of the attack expanded as the days and weeks passed. The attackers went after the high value intellectual property of dozens of U.S.-based companies. It was a style of attack that we have seen before, but that previously had been targeted primarily at governments and defense contractors, not private organizations such as Google.

McAfee was on the front lines of investigating the attack and also in deploying protection. We were able to do groundbreaking research because of the broad scope of our research that covers the full spectrum of security. For example, McAfee discovered that a previously unknown Internet Explorer vulnerability was exploited in the attack on Google and others. Within a day, the attack code that exploited the vulnerability in Internet Explorer became publicly available. This dramatically increased the risk to any IE user. Fortunately, help was available to shield companies against the attacks that used the exploit.

What have we learned from Operation Aurora and from the 4th of July attacks last year? The key is narrowing the protection gap. Traditionally, the protection gap has been up to a week, but with the Korean attack, for example, it is now possible to offer day zero protection against certain attacks.

As an industry, we need to move beyond "defense-in-depth" to "intelligence-in-depth" by tying products together and sharing threat intelligence across them while providing real-time visibility through a security management platform.

So, what does Global Threat Intelligence mean and what should you as a customer look for in intelligence provided by a vendor? Threat Intelligence needs to be real time, reputation-based and interlocked.

This comes down to one thing: There is no point to security anymore. That's as in "point products."

What is the implication of this? As I mentioned, "defense-in-depth" doesn't equal "intelligence-in-depth." We need open security, connecting all vectors and also bringing in core partners, breaking the traditional silos. Only by doing so can enterprises achieve higher protection at a lower cost and move beyond the traditional disparate defenses that form a broken safety net made up of point tools.

....

About the Dave DeWalt

David DeWalt has more than 20 years of experience building innovative, industry-leading technology companies. Since joining McAfee in 2007, DeWalt has guided the company through multiple consecutive quarters of double-digit sales growth, with a record $1.6 billion in revenue in 2008.

DeWalt has a solid record of leading revenue growth at startup, midsize, and large companies, as well as overseeing more than 50 successful mergers and acquisitions. At McAfee, he has spearheaded the company's investment in technology companies. He has also expanded the company's partnerships with leading technology enterprises and directed the growth of the McAfee® Security Innovation Alliance, which now has more than 80 partners.

A recognized authority on cybersecurity, DeWalt spoke before the World Economic Forum in Davos in 2009 about global security threats. He also delivered keynote addresses at the 2009 RSA Conference, the digital security industry's largest annual gathering, and at Interop and Software 2008, two leading technology conferences.

SECURE

Discussions with Senior Management on Information Security

Roseann Larson

Vice President of Global Information Systems,
Risk Management and Compliance – Estée Lauder
Headquarters: New York, NY

The Estée Lauder Companies Inc. is one of the world's leading manufacturers and marketers of quality skin care, makeup, fragrance and hair care products. The Company's products are sold in over 150 countries and territories under the following brand names: Estée Lauder, Aramis, Clinique, Prescriptives, Lab Series, Origins, M-A-C, Bobbi Brown, Tommy Hilfiger, Kiton, La Mer, Donna Karan, Aveda, Jo Malone, Bumble and bumble, Darphin, Michael Kors, American Beauty, Flirt!, GoodSkin Labs, Grassroots Research Labs, Sean John, Missoni, Daisy Fuentes, Tom Ford, Coach, Ojon and Smashbox.

....

Those of us working in Information Security, Compliance and Risk Management share a common challenge: How do we deliver a meaningful, concise message to senior management and the Board on a topic often very technical, detailed and, let's face it, not too exciting? That is, until something goes wrong.

When dealing with these topics, senior management and Board members are our greatest allies and want to understand some key things with their limited time:

- What is the core issue?
- What are the complications?

- *What is it going to cost?*
- *What is the timeframe?*
- *What do you need from them?*

Being prepared to answer these simple questions with a "less is more" approach will be appreciated, as these individuals have many other issues to consider which more clearly deliver bottom line results. While security and compliance may be exciting to us, to the broader organization, it may not be quite as interesting as the upcoming merger, opening a new Asian market or an innovative new product/service. Once the initial presentation is made, it is essential to provide crisp progress reporting on a regular basis.

ADDRESSING THE CORE ISSUE(S)

When answering this question, it is important to tie the issue back to how it impacts the business and strategy - not how many SQL injection issues were found in web code. Technical jargon, acronyms, sexy tech speak are all non starters. Explaining how the issue can disrupt web orders by X dollars is understood by everyone. Describing what other organizations your size are experiencing/doing is very helpful. Sharing Wall Street Journal articles or a Harvard Business Review article for them to read at their leisure can help them get a comfort level with the topic on their own terms.

EXPRESSING THE COMPLICATIONS

Inevitably, the IT Security, Compliance and Risk management topic is full of complications and wrinkles. Senior management and/board meetings are not the place to enumerate each one. Raising the conversation to a high level dialogue focused on the most critical three complications will likely suffice. Answers directed at lack of strategy, lack of historical investment, changing regulatory landscape are things that can be understood readily. Being prepared to answer deeper dive inquiries is essential, but let senior management and/or the Board ask if they need to. Striving to avoid technical diagrams and eye chart slides is also a good goal.

DEALING WITH COST

In these economic times, it is more important than ever to respect cultural norms, budget guidance and internal rate of return targets. Where possible, quantify these amounts in a summary manner, breaking out the year by year budgets, Capital and Expense breakdowns and where possible, rate of return calculations. Often the need for Return on Investment (ROI) calculations can be negated if the investment is directly attributable to a regulation. Becoming expert at identifying ROI opportunities is important as many investments are required as part of holistic compliance/security/risk reduction strategy and as such, may not be justifiable in isolation. Quantifying risk reduction in dollar ranges and using well known formulas may help mitigate cost concerns.

EXPRESSING TIMEFRAME

Straight forward timelines that call out key milestones and highlight key dates seem to work well. Depending on the culture of the organization, more detailed project management charts may be a better fit. Knowing your audience and respecting culture and norms is imperative. When reporting progress, referring back to where the effort started, where it is currently and where it needs to be against a timeline is a good way to answer questions before they are even asked.

ANSWERING "WHAT DO YOU NEED FROM THEM"

It is so crucial to be ready to answer this with actionable items for senior management to own. Senior management can show support in many ways, some more useful than others. Approving budget would seem like the obvious answer, but there is much more.

Giving the IT Security and Compliance function a clear charter and authority to execute are key. Senior management can easily connect key people in their management chain with the CISO, so as to properly educate both sides on strategy and underscore the shared purpose. Associating bonus dollars with compliance/security objectives can be extremely helpful in motivating others in their chain of command. Communications cascaded through the management chains are beneficial to underscore the organizations' commitment to security and compliance.

Senior management can facilitate becoming a part of the formal governance process. Having a presence in the corporate governance bodies helps assure new business initiatives maintain alignment with security and compliance strategy. Additionally, having a seat at the governance table may avoid costly remediation in the future. The business is better informed about the risk associated with their decisions and may decide the risk is outweighed by the benefit derived. Mission accomplished.

Lastly, senior management can meet with you on a regular basis to be updated on progress, new issues and roadblocks. These meetings are extremely useful in that they provide an opportunity to keep the dialogue going. Security and compliance is an ongoing journey and not a snapshot in time. Ongoing reporting of key metrics will allow management to get a better sense of when things are changing.

Senior management and the Board today have an overwhelming number of demands on their time. Viewing security and compliance issues through their eyes, points to reducing risk to the organization and shareholders of security and compliance violations affecting the business. Answering the five questions discussed above will position the CISO as a valuable contributor and will garner continued support and credibility for the larger IT organization.

About Roseann Larson

Roseann Larson is Vice President of Global IT Risk Management and Compliance for the Estée Lauder Companies. Ms. Larson has held a number of positions in her eighteen years with Estée Lauder, spanning from global auditing, new business system architecture, international consumer technology design and development, and most recently focused on IT security, risk management and compliance.

Ms. Larson earned her BBA from Bernard Baruch College. She is a Certified Public Accountant, Certified Information Systems Auditor, and Certified Information Security Manager in the State of New York. She is also a member of the AICPA, NYS Society of CPA's, Computer Security Institute, ISACA, PCI Security Standards Council, ARMA, and the IAPP.

A frequent speaker on the topics of IT security, compliance and identity theft, Ms. Larson has presented before numerous professional audiences around the world. She is an active member of the community, and a graduate of the Huntington Leadership program.

Application Delivery Networks: The New Imperative for IT Visibility, Acceleration and Security

Blue Coat Systems

Headquarters: Sunnyvale, CA
Security Leaders Summit Northeast 2010 – Bronze Sponsor

Blue Coat Systems offers an Application Delivery Network infrastructure that optimizes and secures the flow of information to any user, on any network, anywhere. Blue Coat enables the enterprise to tightly align network investments with business requirements, speed decision making and secure business applications for a long-term competitive advantage.

....

APPLICATION DELIVERY NETWORKS: THE NEW IMPERATIVE FOR IT VISIBILITY, ACCELERATION AND SECURITY

Your business depends on one thing: getting the right information to the right people at the right time. To gain a competitive advantage, improve customer service and access more business intelligence, you have to ensure fast and responsive application delivery to users wherever they are. At the same time, you have to protect users, systems and applications from malicious threats so workers, customers and partners can confidently transact business whether they're using Oracle, Salesforce.com or VoIP.

The key trends driving business today – centralization, mobilization and globalization – often make it difficult, if not downright impossible, to support on-demand application delivery. IT initiatives such as server consolidation and voice, video and data convergence can disrupt network service. Your mobile applications and devices can be compromised by security breaches and data theft. And global IT infrastructures often harbor data silos that are difficult to penetrate and manage, obscuring the view of your IT resources.

If you're not supporting the information delivery your business demands, you need to look at the technologies you're using to monitor application performance, optimize your WAN and secure your Web gateway. Together, these capabilities are absolutely vital to meeting business needs today, and preparing for future requirements down the road.

MANAGING NEW BUSINESS DRIVERS DEMANDS A NEW APPROACH

Businesses that drive toward greater IT centralization, mobilization and globalization need to:

- *See their IT and business processes more clearly.*
- *Accelerate information and business results across the enterprise.*
- *Secure data and users wherever they are.*

To achieve these goals you first need to understand how these business drivers impact your infrastructure.

CENTRALIZATION: THE CHALLENGE OF ACHIEVING GREATER CONTROL AND FLEXIBILITY

Enterprises like yours are always looking for new ways to centralize resources to gain greater IT control and reduce management costs. The more your infrastructure expands, the more you need to consolidate so you can:

- *Enable branch office server consolidation.*
- *Ensure business continuity in the event of a major IT failure.*
- *Accelerate backup and recovery operations.*
- *Regain control of WAN and Internet gateway connections.*
- *Increase flexibility for innovative Web technologies.*

But tightening the reins on your information resources can have a downside, too. Centralization can degrade application performance for branch office employees, making it harder to move quickly in a changing business climate. Strict policy controls can occasionally prevent users from accessing Web 2.0 applications and mash-ups that are used for legitimate business purposes. So the challenge for IT is to achieve and maintain centralized control, while still providing flexible application delivery across the enterprise.

MOBILIZATION: ACCELERATE APPLICATIONS AND DATA ANY TIME, ANYWHERE

Any business that really wants to compete in today's economy has to prepare for a business future driven by ubiquitous connectivity and seamless access to information. It's not enough to drive people to Web sites. Your business has to use technology in smart and effective ways that enable connectivity from almost anywhere. And it's up to IT to help business leaders understand how new mobile technologies can directly support business goals, such as increasing revenue and customer satisfaction.

Ultimately, the goal of mobilization is to improve business productivity and:

- *Accelerate business processes and communication.*
- *Enable key corporate initiatives, such as video for training, VoIP and new application rollouts.*

SECURE

- *Provide the security and performance of a headquarters work experience anywhere.*

Of course, accelerating information to all parts of the world carries significant security risks. To successfully manage on-demand information delivery, you need the intelligence to distinguish business-critical from harmful traffic, prevent data breaches and ensure your business stays compliant with government regulations. You also have to manage the increased strain on your network from Web 2.0 applications, video sites and mash-ups.

Unfortunately, the answer isn't as simple as blocking traffic to questionable sites and content. You need effective ways to regulate the demands on bandwidth generated by personal Internet use without completely denying access. The job of IT is to know how to leverage mobile technologies and Web applications to drive key corporate initiatives while maintaining the security and integrity of sensitive business data.

GLOBALIZATION: SECURELY SCALE YOUR BUSINESS TO LEVERAGE NEW GROWTH OPPORTUNITIES

Globalization is one of the most important business drivers today. Any company that wants to succeed globally needs the right mix of tactical and visionary IT leadership to handle everyday management issues while strategically preparing for future growth opportunities. The more your business expands across the world, the more your IT organization needs to:

- *Mitigate risk and ensure compliance through integrated network security controls based on visibility, intelligence and policy.*
- *Change policies on demand to instantly respond to evolving technology, organizational and market conditions.*
- *Prevent data leaks with centralized policy controls over network use and access.*

- *Proactively respond to application performance demands across a growing distributed enterprise.*

Today, for instance, global transactions increasingly depend on softwareas-a-service (SaaS) applications such as Salesforce.com, and unified communications such as video conferencing and instant messaging. In addition, service-oriented architectures (SOA) combine applications and data from multiple sources, so your infrastructure must understand what is business-critical and what isn't. You have to accelerate essential applications, stop unwanted or dangerous traffic and manage or mitigate everything in between.

In addition to optimizing daily network performance, many companies are already preparing to deliver the next wave of anytime, anywhere business applications by converging their voice, video and data networks. The business benefits of network convergence are clear: fast, dependable, real-time communication, unprecedented information mobility and worker flexibility, and the ability to instantly deliver a variety of data types to almost any device, anywhere in the world. But how can you ensure these new models of application delivery enforce your corporate security policies, mitigate security risks and help you achieve compliance?

And, more importantly, what will information delivery look like in the next few years, and how should you prepare for it now? These are questions IT leaders need to address to help the business maintain a competitive edge in the global marketplace.

THE FUTURE OF APPLICATION DELIVERY

While the challenges of consolidating and mobilizing global IT resources are far from simple, you can start to tackle them by learning how to achieve greater application visibility, acceleration and security within your infrastructure.

One of the biggest obstacles to delivering fast, secure and reliable application performance lies in the connectivity layer. While networks route traffic and deliver packets very efficiently, the connectivity layer can't tell you whether the content is helpful, harmful or even potentially catastrophic. To adequately assess the quality of network transactions, application performance and user experiences across your network, you need a new layer of IT control that helps you confidently deliver applications, not just packets.

THE APPLICATION DELIVERY NETWORK: A NEW LAYER OF INTELLIGENT CONTROL

While many companies can manage the individual issues that compromise application performance, every enterprise still needs a critical layer that can go beyond just keeping applications up and running. You need the ability to stop malicious activities and deliver applications precisely when and where they're needed most. Now and in the future.

The Application Delivery Network does just that. It answers the demand for greater application mobility and security in a changing global business environment. By combining three core capabilities – Application Performance Monitoring, WAN Optimization and Secure Web Gateway technologies – the Application Delivery Network helps you:

- *See applications and users and how they behave on the network, and troubleshoot performance issues.*

- *Accelerate mission-critical applications, streaming video, SSL and other enterprise applications.*

- *Secure against malware, data leaks and performance degradation.*

- *Enable a highly efficient and productive end-to-end user experience anytime, anywhere.*

BUILDING THE APPLICATION DELIVERY NETWORK

To help you get started, this white paper describes the components and processes you need to establish an Application Delivery Network in your infrastructure.

APPLICATION PERFORMANCE MONITORING:

Identify and resolve the critical factors that impact performance.

Monitoring application performance is one of the toughest and most essential jobs in any IT organization. To do it well, you have to:

See all the traffic on your network.

This kind of intelligence work requires tools that can automatically identify hundreds of applications on your network every day. To separate the good from the bad, you have to distinguish between business-critical content and malware and recreational traffic such as iTunes, YouTube and peer-to-peer (P2P) applications. You have to be able to sub-classify complex environments, including Web-delivered suites such as Oracle and SAP, so the most important operations and users within these applications have the highest priority.

Monitor the end-user experience.

When your IT organization is successful, users rarely know how hard you work to keep applications up and running, maintain data integrity and prevent hardware crashes. But preventative measures are only as effective as the tools in your arsenal. You also need the ability to measure and alert IT about specific factors that impact the user experience, such as P2P applications that steal bandwidth from your critical business applications.

Determine the source of network problems.

Application performance problems can come from almost anywhere, so you need to execute a range of tasks to quickly isolate the main issues, whether it's a delay between the server and network, or problematic hosts, servers or applications that cause the most performance degradation. Next you have to analyze the causes, whether it's a spike in application usage or a protocol problem, and use this detailed analysis to fine-tune the environment and get your applications back up to speed.

Get end-to-end visibility.

Many IT organizations have traditionally performed these tasks with a diverse set of connectivity layer tools from different vendors, which can lead to compatibility problems. However, a complete, integrated solution is more scalable and cost-effective in the long run, and can help you achieve ROI faster than a disparate group of tools that require integration. Application acceleration and control technologies also integrate service-level metrics and statistics into comprehensive reports that help you manage the user experience by identifying and resolving problems quickly.

WAN OPTIMIZATION:

Outstanding business performance depends on fast application performance. Manage the user experience throughout your distributed enterprise.

Most corporate environments today have hundreds of applications running on the network at any given time, in any location. They can all have different management issues, depending on the application type:

- **Internal bulk applications**

    This type of application uses CIFS, MAPI, FTP or TCP and includes file access, storage consolidation, email, Internet, backup and applications used to distribute updates, patches and new firmware images around your company.

- **External applications**

    These applications cross the spectrum from business-critical to downright malicious. Therefore, external applications can be the most challenging – and the most important – to manage. They can include everything from corporate training videos and Salesforce.com to the hundreds of recreational applications your users access from your corporate network.

    Managing access to recreational sites can be tricky, because many of these sites, such as YouTube or instant messaging, are sometimes used for business purposes as well as personal entertainment. So the key is to distinguish between business and recreational use, maintain control of your network resources and stop the malicious traffic that can piggyback across your network.

- **Real-time applications**

    Applications such as VoIP, video conferencing, credit card transactions, financial trades and order processing are often the most sensitive to delay and the most critical to the business. How can you keep them available 24/7?

- **Centralized management and control from data center to device.**

    To help ensure consistent performance across all these application types, you have to manage and protect business data all the way to the device level. And that requires the ability to enforce corporate policies and protect information from theft, loss and malware outbreaks. As a result, you can accelerate the performance of all your client/server applications and enable transparent access among VPN users who occasionally work in the office.

SECURE WEB GATEWAY:

Anticipate and block threats to your business at the point of entry. You have to be ready for anything.

You never know which new form of malware will try to breach your gateway to steal or alter financial data, personnel information, customer histories and more. That's why your technology strategy must constantly evolve in response to potential threats, both physical and digital. The challenge for large enterprises is to ensure global security without compromising global agility. To secure your entire enterprise, you have to protect every point of access to your business – the devices and workers themselves.

Any IT strategy should include these four key security requirements:

- *Protect against malware.*

    Technology from leading anti-malware vendors such as Kaspersky, Sophos, Panda and McAfee can help prevent malware outbreaks by filtering both incoming and outgoing Web traffic in real time, and reduce your exposure to malicious Web content.

- *Guard employee productivity.*

    Not all recreational applications are counterproductive to your business. You just need the right tools to know when recreational use is depleting bandwidth and employee productivity, so you can control it according to your corporate policy and business needs.

- *Prevent information leaks.*

    Ensure your data leakage protection capabilities offer the ability to watch, alert and prevent the theft of information from databases and other vulnerable resources throughout your company. Consider tools that integrate with Symantec/Vontu, RSA/Tablus, Code Green Networks and other leading vendors.

- *Validate trust.*

  The bottom line is, you simply have to secure your enterprise from the inside out, because business loss comes in many forms: information theft, downtime, decreased productivity, accidental or intentional data corruption and more. With all these threats, it's easy to wonder if you'll ever have enough security to protect your business. But by integrating extensive capabilities for URL filtering, data leakage prevention, malware protection, policy management and identity authentication into your network management arsenal, you can dramatically improve your overall security approach.

....

About Blue Coat Systems

Blue Coat provides the intelligent control you need

You face a converging set of business drivers – centralization, mobilization and globalization. While many companies provide the tools to manage aspects of these business challenges, only Blue Coat helps you tackle them head-on with Application Delivery Network solutions. Our core technologies – application performance monitoring, WAN optimization and Secure Web Gateway – deliver the visibility and control you need to see, secure and accelerate applications like never before.

Visibility

Blue Coat Application Delivery Network solutions provide the ability to identify and classify applications and users across the network. Our capabilities allow you to discover all application traffic, monitor the user experience, troubleshoot performance issues and resolve problems before they impact the user experience. More specifically, we help you to:

- *Automatically discover 600+ applications*

- *Identify P2P, recreational or streaming applications over any port*

- *Sub-classify complex applications such as SAP, Oracle, Citrix, Web, CIFS, MAPI and DCOM*

SECURE

- *Discover URL and external sites within HTTP*
- *Identify problem hosts, servers and applications*

Acceleration

Blue Coat helps you accelerate business-critical applications, including internal, external and real-time applications to any user, anywhere. And we do it while ensuring a headquarters work experience, wherever your users are located. Our acceleration technologies include:

- *Object and byte caching*
- *Compression and basic quality-of-service (QoS) capabilities*
- *External Web/SSL acceleration*
- *Protocol acceleration for TCP, CIFS/NFS, MAPI, HTTP and more*
- *Advanced Web policy and bandwidth management*
- *Advanced application ID technology*

Security

Blue Coat secures your Internet gateway to help protect users from malicious content and applications. Our security capabilities include:

- *Anti-virus and malware scanning*
- *URL and Web content filtering*
- *Centrally managed distributed gateway*
- *Granular policy management across over 500 variables, including user, group, application, source, content types and transaction*
- *Logging, statistics and SNMP support*

Blue Coat Application Delivery Network solutions deliver the intelligent control you *need centralize, mobilize and globalize your entire IT infrastructure. With Blue Coat, you can optimize application and network performance for any user, anywhere across a distributed enterprise. Let your IT organization become the engine that drives greater business efficiency, effectiveness and competitiveness. Learn how implementing an Application Delivery Network can help your organization prepare for the next wave of converging IT and business challenges. Find out more at www.bluecoat.com.*

SECURE

Trends – Insights into Preparing for the Future

John Walp
Administrative Vice President and CISO – M&T Bank
Headquarters: Buffalo, NY

M&T Bank is considered one of the country's most highly regarded regional banks. We were founded more than 150 years ago in Western New York, where we are still headquartered. Our parent company, M&T Bank Corporation with current assets of $68.2 billion (as of June 30, 2010) is one of the 20 largest commercial bank holding companies headquartered in the U.S. Our customers are able to bank at more than 750 branches throughout New York, Maryland, North Carolina, Ohio, Pennsylvania, Washington, D.C., Virginia, West Virginia, New Jersey, and Delaware and be served by friendly and professional staff. In addition, our customers have access to a sizable ATM network of more than 1,800 locations as well as state-of-the-art telephone and Web Banking. M&T Bank is recognized for its financial strength and sound management. This strength, along with our entrepreneurial philosophy, has made it possible for us to pursue a course of well-managed growth.

....

We are living in truly an incredible time in history where the Internet has become woven into almost every aspect of many people's daily lives. With unprecedented access to information, content, and the things our children take for granted such as iPods and YouTube, technology would certainly shock and amaze the average American living in 1985, the year I graduated high school.

The Internet has fundamentally redefined many areas of the American economy and in some instances, eliminated entire industries. It has also given birth to entirely new industries; one's which offer great promise, but also introduce risk in ways that are often hard to quantify.

As this rapid inter-connection and inter-dependencies of everything from personal computers to cell-phones has occurred, so has the threat to the highway that carries the 1's and 0's that make up all digital communications.

To deal with this threat, Information Security or Information Assurance professionals have almost always taken the approach of building security models dating back to the Dark Ages. The "castle and moat" approach to securing digital assets proved adequate for the Internet and the businesses leveraging it in the late 1990's and early 2000's. But something drastic has happened in the last 2 to 3 years as the explosive growth of technologies such as FaceBook and Twitter have challenged the time tested approaches to protecting business information.

No longer can a large corporation depend on the aspect of perimeter defense, as we've seen some of the greatest risks to business's most valuable data, have come from the enemy within. International Organized Crime and nation-state sponsored hackers have leveraged the Internet to routinely raid our nation's largest corporations and pilfer terabytes of intellectual property. The threat from malware such as Zeus has undermined our confidence in tried and true security controls such as anti-virus and two-factor authentication. Consumers and business have recorded record losses from Internet crime and there seems to be no end in sight to the speed and complexity of today's attacks. While information security professionals have done their best to defend against this growing cyber threat wave, I am afraid we are only treading water as a profession and may indeed drown under the crushing flow of attacks, if a new approach to protection information is not found. This is not meant to undermine some of the best methods of protecting information such as encryption, but more rather focused on the aspect of, if we are trying to protect everything, we protect nothing.

The castle and moat approach has caused information security professionals to think about the challenge of protecting critical information in the wrong way. As we have seen demonstrated by the Zeus malware attacks, the criminals don't even have to bother trying to crack the defenses of the "castle", rather they can focus their attention on the millions of unprotected "villagers" whose personal computers have become all too easy targets. Millions of personal computers are routinely compromised and become part of massive zombie "botnet" armies as was revealed in the Mariposa Botnet in which 12 million computers were under the control of three cyber criminals operating out of Spain.

Even when the "villager's" PC is not the target, we've seen extremely sophisticated and lengthy attacks where the cybercriminals operate with complete ambiguity and can fly silent beneath the radar of information security systems such as intrusion detection. As Albert Gonzales demonstrated in his compromise of Heartland Payment Systems and other payment card processing corporations, our adversaries are patient, intelligent, well funded and collaborative in order to overcome hurdles as part of their operations. The Royal Bank of Scotland WorldPay hack was both brilliant in design and execution. The result was a $9 million loss in less than 12 hours involving hundreds of accomplices.

The picture I paint may seem grim and hopeless to some, but I believe that there is a silver lining in this dark gray cloud of cybercrime. There is an opportunity to look at the business of protecting information in new ways which take lessons from a long history of criminal activity. We must understand our adversaries and the motives behind the crimes they commit. We must deeply understand the way the criminals collaborate and share information. Only then can we begin to design ways to detect, defend and defeat this cyber menace and ensure the integrity of a truly world changing technology like the Internet does not dissolve into an un-trusted medium for business transactions.

As we look into the future, the growth of the Internet will be unrelenting as more and more devices become connected. The deployment of IPV6 will allow the unrestricted growth of the Internet so our cars, smart phones and other devices in our home such as refrigerators and televisions can be connected. Information content delivery will continue to evolve as devices such as the iPad and companies such as NetFlix, Hulu, and Pandora, redefine how people consume content and information. The expectation of anywhere access will be common place as the explosive growth of smart phones and other mobile devices allow access to information while in the office and at the beach. Location-based services will challenge privacy as a generation of users describe their daily lives in near real time online.

All this will drive most of our old security models and controls used to protect information, the way of the dinosaur. The ever growing volume of log data and ability to access information from anywhere on the globe, will drive security product companies to build solutions which provide the ability to analyze user and network behavior in ways that allow proactive anomaly detection. Some may argue these solutions are available today but the software that performs this analysis today is not "smart" and is only as effective as the quality of the data and people who manage the system. I believe there will be a new class of solutions that use yet to-be defined internal standards for data and log ingestion, similar to how the payment industry has, in some ways, become globally accepted.

The focus of this new security way of thinking must be centered on the data as it is at the heart of the castle. By linking the business value of data to the necessary security controls, we can rely less on the "blocking and tackling" technologies such as firewalls and intrusion detection/prevention systems. Encryption must be commoditized and deployment simplified to deal with the legacy systems running much of our core national critical infrastructure. Only then can we protect our most valuable digital assets and move away from the idea of security being our borders with the technical castle wall. Data is king and key to every business process and is in the sights of cyber criminals' virtual guns used to commit their crimes.

Developing new ways to protect our most valuable assets is the only solution to the growing threat of cyber crime and never ending cycle of vulnerability and patch management. We must instill cyber security into the very fabric of our society and think of it in much the same way we think of a car seatbelt safety. It is a risk decision not to wear a seat belt; however, it is against the law in nearly every state in our nation. Seatbelts save lives and can be proven with mathematical certainty. However, some still choose not to take advantage of the vital safety and risk reduction feature; the consequences of which can both be life changing and financial in a variety of ways.

The recognition of cyber security as a national security issue demonstrates the urgent need to not solve the problem after the "horse has left the barn." Our children must learn about safe online practices from the very first moment of their exposure to computers and the internet. Our higher educational institutions must weave cyber security into every applicable curriculum. This will ensure we do not continue to produce cyber security ignorant workers and citizens that are unaware of the threats they face and provide endless opportunities for our adversaries and criminals to exploit.

About John Walp

John Walp has more than 18 years of Information Technology experience, more than half of which has been focused on information security challenges. He currently serves as Administrative Vice President and CISO for M&T Bank, a $68 billion financial institution headquartered in Buffalo, NY. Previously, he held the role of Vice President, Network Security Solutions Manager for M&T.

John is a Certified Information Systems Security Professional (CISSP) as well as a Certified Information Security Manager (CISM). He is a graduate of the FBI Citizens Academy and serves as the Executive Vice President of the FBI's Buffalo InfraGard Membership Alliance. Mr. Walp serves on the advisory board for the Center of Excellence in Information Systems Assurance Research and Education (CEISARE) at the University of Buffalo as well as the Business Advisory Board of Medaille College.

He is a member of the High-Tech Crime Consortium and the U.S. Secret Services Electronic Crimes Task Force.

A Veteran of the United States Air Force, he served his country for 22 years which included both active and reserve duty service. In 2004, Mr. Walp was recalled to active duty and deployed to the State of Kuwait in support of Operation Iraqi Freedom and Operation Enduring Freedom. He was selected as part of an elite logistics cadre to aid in establishing the U.S. Central Command's Deployment and Distribution Operations Center.

He holds a Bachelor of Science in Computer Information Systems from State University of New York College at Buffalo. He and his wife Laurie have four children and make their home in Amherst, NY.

SECURE

Why Enterprises Need a Flexible Approach to Data Security

Sean Glynn

Vice President, Marketing – CREDANT Technologies
Headquarters: Addison, TX
Security Leaders Summit Northeast 2010 – Bronze Sponsor

CREDANT Technologies offers the flexibility to choose the encryption solution that best meets your data protection and compliance needs, delivering data encryption across any endpoint - desktops, laptops, handheld devices and removable media - including our patented, intelligent data encryption solutions as well as new hardware and software based full-disk encryption offerings.

....

We all love simplicity. It's why when we make our shopping decisions, we often tend to prefer supermarkets or shopping malls where everything is under one roof. But there's a downside to this: While it offers the advantage of convenience, customers could be missing out on the expert advice, better quality products, and personalized, individual service available from more specialty retailers.

Enterprises love simplicity too. That's why so many have a single point of contact for their IT needs, that's why so many choose equipment from a narrow range of suppliers. Going back in time, it's the reason why Microsoft made such headway in the workplace – customers loved the fact that they could pick up a machine from any vendor and make it work with a machine from another vendor.

There is however, a downside here too. Enterprises comprise a complex web of individuals all with differing needs and using multiplicity of devices for communication. And while simplicity sounds fine in principle, this organizational complexity can lead to problems of its own.

The inherent complexity of organizations is often ignored when it comes to the procurement of security products. Enterprises will hone in on a single supplier – often times a broad-based security vendor, where there is interest in that vendor's products. There are plenty of these in the market, all with a broad range of products and purporting to have an integrated approach to security. But it's not all that simple!

Of course, most customers realize that given the broad product range offered by these companies, they might not always have the most perfect solutions in every case. But while they might not produce the best anti-virus product or the best anti-spam products, the products are relatively easy to install and maintain. One high profile example of this drive to simplicity we've seen just recently was the news that Intel is to buy security software company McAfee … Intel has promised to integrate McAfee with its chip technology, implying that we can expect in the near future that PCs will be secured from the outset.

It might seem on the face of it to be a simple approach, but I worry that in the real world, it's not going to be good enough! The modern enterprise is under threat from a variety of different sources, by a variety of different methods and companies should be wary of the fact that a single supplier may seem like the simplest way, but is not always the best way.

The same process applies with choosing encryption products. If you don't go to a specialist encryption company who can understand your requirements, and offer expert guidance on the best mix of technological solutions to meet your business needs, then the obvious move is to go to a more general security vendor who will offer you a single type of encryption product. Sometimes, it will be Full Disc Encryption (FDE), sometimes it will be all self-encrypting disks (SED) or sometimes it will be policy-based encryption. Whichever it is, the customer is not buying according to their particular needs or environments – but rather, buying the product their vendor wants to sell them. The reality though is that this approach is fraught with risk as distinct user requirements often call for different types of encryption product, and user organizations should be considering multiple technologies to determine how best they can enable their users for optimum performance while mitigating their risk and protecting their data.

For example, organizations might determine that certain users or groups need the latest generation of self-encrypting hard drives on their laptops, to maximize performance – while others will require a solution suitable for multi-user desktops or laptops – and many companies will need an external media encryption solution that offers additional protection for mobile workers or people handling particularly sensitive information. Enterprises don't want to be reading newspaper stories about unprotected USB sticks being left in clothes at the laundry, or laptops turning up in the back seats of taxis.

Many organizations for instance are currently considering the encryption offered by Microsoft's BitLocker as being a potentially adequate solution for their needs. When it was launched, Microsoft made much of BitLocker, the encryption technology included in Ultimate and Enterprise versions of Vista and Windows 7 and which offers full-drive encryption. Microsoft makes much of the fact that BitLocker is "good enough" encryption for many organizations, but the fact is that while it's a useful point of call, BitLocker is not going to be the answer for every company or for every user. And, while it's a useful starting point, there's a requirement to manage BitLocker, just as with any other encryption technology.

That's not all: by actively pursuing the one-size-fits-all approach, companies run the risk of failure in other IT deployments or processes. In other words, by trying to squeeze a particular technology into a corporate IT environment, you run the risk of breaking another part of the system.

But while it's clear that adopting the one-size-fits-all approach isn't necessarily the right way, there is one advantage in adopting this approach – it means that there's a single management framework to worry about. And management is a big headache for companies – again, the simpler the better.

This is where Credant comes in. The company makes no claim to be an all-purpose IT company or even an all-purpose data security company, its expertise is in encryption and data protection, and nothing else. But because of this expertise, it can provide products using all types of encryption, with technologies specifically designed for particular environments or to support particular security policies. What's more the company has the expertise to realize what the areas of difficulty are and what policies need to be implanted.

Credant has realized that companies that have tried to deploy a single encryption technology as a means of solving their endpoint data protection needs have often created a whole new range of complexities across their organizations. Credant will help these companies set common data protection policies when multiple encryption technologies are being used, all the while building forward compatibility to deploy and manage new encryption technologies that come onto the market.

What does Credant offer its customers? Most of all, Credant helps customers adopt a flexible approach to data protection and risk mitigation. For one thing, Credant is the largest endpoint data protection specialist in the world and is the only company that supports all endpoint encryption approaches: policy-based, FDE, SED as well as offering the world's best removable media encryption product, all handled through a central management framework. It is this range of different solutions and the breadth of expertise that makes the company the most attractive port-of-call when looking for an effective company-wide endpoint encryption solution.

With the growing number of security breaches and the increasing number of threats, companies have to take protection of their data very seriously and seek out the best possible security products. The right encryption products are part and parcel of that approach. While a one-size fits all approach is superficially attractive, it's not going to be the solution of the enterprise that takes data protection and encryption seriously. Simplicity is not always the best answer.

credant.com

About Sean Glynn

Sean Glynn brings nearly 20 years of international IT and Security experience to his role as CREDANT's Vice President of Marketing. Prior to his current role at Credant Technologies, Sean has worked in North America and in Europe - with companies like Sophos, McAfee, BMC Software, Compaq and Dell Computer – developing and bringing to market security solutions to meet customer data protection and compliance needs.

Acceleration of Change in Information Security

"An essay in which the rules of proof and logic are blatantly subverted to support the author's pre-determined conclusions"

Roy L. Post

Chief Information Security Officer – AXA Equitable
Headquarters: New York, NY

In business since 1859, AXA Equitable Life Insurance Company (formerly The Equitable Life Assurance Society of the United States) is a leading financial protection company and a premier provider of life insurance, annuity, and investment products and services. AXA Equitable's assets under management totaled $519 billion as of June 30, 2009.

....

It goes without saying that security professionals will tell you that the demands on information security are increasing. I claim that there is a well known framework of facts in support of that assertion.

- **Moore's "law"** – The power of a computer chip (processor speed, memory capacity) doubles every 18 months (Gordon Moore, founder, Intel)
- **Gilder's "law"** – Network bandwidth doubles every 6 months (George Gilder, "Telecosm")

Moore multiplied by Gilder is giving us smaller, more powerful, more portable and more interconnected information devices. Some of that is reflected in accelerating patent statistics, for example.

- *Patents*
    - *In 1981, 65,771 US patents were issued*
    - *In 1995, 101,419 US patents were issued – a 54% increase in 15 years*
    - *In 2009, 167,439 US patents were issued – a 65% increase in 15 years (US Patent and Trademark Office, "US Patent Activity, 1790 to present")*

INTERNET USE STATISTICS SPEAK FOR THEMSELVES

Worldwide, the number of internet users increased 182% from 2000 to 2005, approx 30% per year. Now, there over 2 billion internet users. (internetworldstats.com) We still haven't connected everyone to the internet and there's still a lot of room to expand. The world population is about 7 billion people right now, growing at 1-2% per year (U.N "World Population to 2300", 2004)

So what happens when all those users are on the same network with "anybody to anybody" connectivity?

- *Metcalfe's "law" – The value of a network grows as the square of the number of users (Dr. Robert Metcalfe, polymath)*

Think "Web 2.0" such as eBay, Wikipedia, StubHub, Craig's List, the iPhone app store and so forth. These are services that become ever more valuable as the number of users increases.

SO, HOW DOES THIS LOOK THROUGH THE LENS OF INFORMATION SECURITY?

We're moving from a near-monoculture of Windows operating systems to a world of many different operating systems. We never completely secured the Windows world. Now we are faced with an explosion of networks, operating systems and applications.

- *Cell phones and smart phones*
    - *Around 4.6 billion cell phones right now (Wikipedia "List of Countries by Mobile Phones In Use")*
    - *Less than 10% of those are smart phones*
    - *Smart phone penetration has a lot of room for growth*
- *Operating Systems*
    - *On desktop and laptop, Microsoft and Apple seem to have settled into a static pattern*
    - *But on smart phones, Wikipedia lists at least 8 operating systems*
    - *Tablets and dedicated readers introduce other OS's as well*

Every smart phone manufacturer must envy the success of Apple in attracting apps. We know that applications in the desktop and laptop world frequently introduce security flaws. It is unlikely that the rapidly growing forest of smart phone apps can be any better.

- *Count of iPhone apps:*
    - *May, 2008 – 1 app*
    - *May, 2009 – 17,000 apps*
    - *May, 2010 – 38,000 apps (source: 148Apps.biz "App Store Metrics")*

WHAT HAS THE SOCIAL RESPONSE TO THIS KIND OF GROWTH BEEN SO FAR?

- *Acts of Congress & other notable events: In the 12 years 1999 – 2010 there were at least 20 significant acts of Congress, individual states and the SEC that significantly impacted Information Security (*see end of chapter for list)*
- *Policy frameworks: Archer Technologies lists over 30 "authoritative sources" for Information Security frameworks, regulations, policies, standards and/or requirements*

The impact of these trends on information security is ubiquitous, running from the full scope of an enterprise down to the individual worker, and from the full breadth of a country down to the individual citizen. Three (among many) overall trends can be isolated:

1. *Consumerization of hardware, software and software services*
2. *Evolution of internal IT departments into integrators of external services*
3. *Increased reliance on legal and regulatory controls over technical controls*

SO WHAT DOES THE INTERACTION OF THESE TRENDS LOOK LIKE?

For one thing, a worker will start seeing targeted advertising while using approved consumer solutions for work purposes. The worker already does today, just by using a search engine to find out how to spell a word. As consumerization proceeds, an internal email about a client routed through consumer email will trigger a tempting ad for a personal cloud-based CRM solution with a one-month free trial. Uploading a spreadsheet into the CRM solution will trigger an ad for new personal hardware with six months of free broadband over a different provider. Using the free broadband service while reviewing a client's portfolio on a financial analysis website will trigger an invitation to apply for a job with a competitor.

The worker will video conference with clients over any channel provider in any location where she spends time – home, office, airport, airplane, restaurant, day-care center, health club, car, taxi... Some of those channels will be secure. Others will not be, and many will be impossible to assess. The individual worker is now in a cultural and technological environment where security controls daily become more difficult to enforce. Education, outreach contact and reinforcement may be the tools with the most lasting impact.

At a rapidly increasing rate, IT services continue to be commoditized and sought outside of the enterprise. IT is morphing into a group whose chief function is as an integrator of services delivered by third parties. Integration will not mean that all services and data-flows somehow come back through the corporate mothership, functioning as a central hub keeping track of what is going where and when. Outside service providers are horizontally connected; for example passing data from a brokerage to a consolidator to a mailing fulfillment center, all operating as separate business entities working on behalf of a client firm.

Thus, one data sharing connection from the corporate mothership to a single service provider implies many more connections and repositories of confidential corporate information outside of the enterprise. Information security risks along the supply chain are on an exponential growth curve. Under this load, security issues may go unobserved and only be handled through contractual management of liability rather than be detected and prevented in the first place. Assignment of liability alone cannot defend against a reputational risk.

Organizations, as a whole, will become impatient with risk management processes. Businesses need risk; without risk there is no growth.

Risk management will have to stay responsive and agile and struggle hard to keep up with the speed of business innovation and change.

Security frameworks are cumbersome and blunt instruments at best, but as risk management tools they prove effective within organizations. Between two or more organizations, though, interpretations and responses to frameworks inevitably differ. When corporate security managers assess another business entity in the services supply chain, those differences degrade trust and lead to increased perceptions of risk.

Information sharing forums in which businesses come together to compare and share best (and worst) practices are essential to enhancing trust and helping security managers spend their time on real instead of misperceived risks. Public-private forums can further enhance that sharing by giving the regulators and legislators the information they need to govern rationally and help business leaders understand and influence government perspective and intent.

In the information security space – a world whipped into frenzy by technological change, regulatory response and corporate strategy – the answers to managing it will not be found in new layers of silicon solutions or legal constructs. Answers will be in the "soft" solutions.

Awareness, training, reinforcement and outreach must continue to help increasingly dispersed and mobile workers navigate the evolving workplace securely.

Gaps in securing the supply chain will best be solved in information sharing forums among like-minded subject matter experts willing to come together and help each other toward a common purpose – aligning their diverse solutions to help secure that frenzied and interconnected world.

* **Significant acts of Congress, individual states and the SEC that impacted InfoSec:**

1999
Gramm Leach Bliley Act

2000
Electronic Signatures and Global Commerce Act

SEC Reg S-P

2001
USA PATRIOT Act

2002
Sarbanes Oxley Act

E-Government Act

Homeland Security Act

2003
Do Not Call Act

FACT Act

CAN SPAM Act

SEC Reg S-P redux

Calif SB-1386

2004
Intelligence Reform and Terrorism Prevention Act

2008
Economic Stimulus Act

Emergency Economic Stimulation Act

2009
American Recovery and Reinvestment Act

Fraud Enforcement and Recovery Act

Credit CARD Act

2010
Dodd Frank Wall Street Reform and Consumer Protection Act

Massachusetts Privacy Law

About Roy Post

Roy has been Chief Information Security Officer for AXA Equitable since 2003. AXA Equitable is one of the U.S. based operating companies of AXA Group, a global insurance and financial services company headquartered in Paris, France.

Prior to joining AXA, Roy was an IT director for Bristol Myers Squibb leading a team of software developers specializing in decision support.

Roy graduated from SUNY New Paltz in 1977 with a BA in Mathematics and Computer Science. Roy holds CISM and CISSP certifications.

SECURE

Strategies for Building a Successful Roadmap

Todd Colvin

Director Enterprise Data Security – Paychex, Inc.
Headquarters: Rochester, NY

Paychex, Inc., together with its subsidiaries, provides payroll, and integrated human resource and employee benefits outsourcing solutions for small to medium-sized businesses.

Imagine if you will, the adventurous life of an ancient mariner sailing the open seas in search of glory for the purpose of conquering or defending some unidentified land mass that no one is really sure exists. To be successful, the mariner must risk life and reputation, by: developing a defensible case for taking on any such adventure, seeking sponsorship from a wealthy financier with undoubtedly unreasonable demands, side-stepping politics while attempting to meet expectations, rigorously researching and forecasting details, plotting an expedient course, stocking a sea-worthy vessel with crew and supplies, and navigating unchartered waters through potentially turbulent weather, before hopefully reaching what could potentially be a landmass populated with hostile natives.

If you were to rough sketch the job description of an ancient mariner by combining all of their unique skills and experience (some of which have been identified above), along with the fact that one miscalculated step anywhere within the process might result in an untimely demise, you might wonder why anyone would ever seek employment in their trade. With little modification to the ancient mariner's job description and skill requirements, other than details regarding relocation benefits, most employers could openly post the same career opportunity today, but under the title of Information Security Leader.

Although briefly advantageous for many employers to take this approach, as most are still uncertain what skills an information security leader should possess, the real advantage here is to the benefit of a skilled information leader capable of developing and executing on a overarching vision of information protection that serves to meet the needs of their employer by advancing business objectives. By tactically using the skills of an ancient mariner to include: envisioning, servitude, cartography, plotting a course, logistics, navigation and error correction, meteorology, and fleeting accomplishment, the information security leader can build a strategic roadmap to surmount nearly any foreseeable condition. But where does an information security leader begin the development of a roadmap? By picturing the destination…

Envisioning is an often overlooked activity that by some arguments is the most critical element in the creation of a roadmap. Imagine if you will that you're an ancient mariner standing by the shores of a vast body of water pondering what is on the other side. Recognizing that you possess insufficient wealth to travel abroad, you must therefore convince others that it is worth their investment in your adventure to explore distant lands. What vision will you paint if given an opportunity to present your idea to a prospective benefactor? Would you paint a vision of your current world view, or would you paint a distinctly exotic vision with the potential for untold riches? If you take the latter path, your sales pitch will be contingent on the depth and breadth of detail that you bring to your vision in combination with your resource requirements, time tables and proposed return on investment.

If you're a security leader, you too need to take time envisioning uncertain future states that may impact the confidentiality, integrity and availability of the information systems and resources that you're chartered to protect. Leaders should understand general trends in people, process and technology. They should have a sense of the growth and consumption of information by today's post-industrial society. Leaders need to look three, five, even ten or more years into the future to understand business, societal and technology changes lurking on the horizon.

If you're a leader having difficulty envisioning the exotic, begin with the tangible. Ancient mariners often sought the counsel of cartographer's (i.e., map makers) with their documented pictorial of the known world. Discussions with a cartographer offered the mariner a perspective on how far they could travel before crossing the Rubicon. In some instances, maps painted an incomplete picture of "known boundaries," thus requiring mariners to seek out additional sources of knowledge. This knowledge could come from the accountings of others having sailed in similar uncharted waters, written collections of exotic destinations, the research of educational institutions, and the predictions of leading authorities on the subject.

As modern leaders, we're bathing in knowledge thanks to seven by twenty-four news reporting, social media, email, IM and the Internet at large. The difficulty though, is whether the available knowledge is purely status quo, group think or a lazy person's enlightenment. Therefore, security leaders must balance the aforementioned modern information traps, as they're beneficial for monitoring the pulse of society in general, with that of past performance and future predictions. Past performance is relevant in that history does repeat itself, so a leader that is looking to and learning from the past may potentially avoid a future recurrence, while at the same time looking ahead to future predictions to detect general patterns of acceptance and adoption for new practices and technologies. For example, could avatars, virtual worlds and game design all collide in the creation of a new and broader market place for commerce that is accessible to all, regardless of technical competency? If so, how have you adjusted your roadmap to address this possible scenario?

So where does an information security practitioner buy their crystal ball? Through participation in industrial security groups, trade magazines, technology centric forums, technology agnostic forums, social sciences, climatic research, sources of social prediction, advanced research on security mechanisms, conspiracy theories and just about any other form of knowledge that a leader can bring to bear when attempting to envision an uncertain future. A leader must adopt the mantra of read, discuss and think. Read about past, present and future trends, discuss them with peers and visionaries, and then think how the results of this knowledge can and should shape a roadmap.

Ancient mariners would consume all available knowledge prior to plotting a course to their proposed destination (i.e., vision) during which time they would incorporate decisions regarding open water currents, trade winds, waypoints, sand bars, rock outcroppings, optimum travel periods, average durations and any contingencies or dependencies that should be calculated into the intended course. They may also factor in lessons learned from the previous attempts of others, changes in societal expectations, benefactor expectations of territorial expansion or wealth accumulation, or the introduction of royal decrees of compliance with new rules and standards. In essence, they built their roadmap factoring in every piece of available knowledge while attempting to identify any uncertainties along the way to minimize loss of life while meeting the goals of their benefactors. They also built their roadmaps to entice investors into supporting their goals to expand and protect the empire from unnecessary harm in support of the expectations of the monarch and their subjects. Information security leaders must follow the lessons of ancient mariners, by placing quill-to-parchment in the creation of a roadmap.

Once a clearly written document outlining your ideal operating state is completed—what you envision—you must then lay out a course of action leading you to your intermediary destination (more on intermediary destinations later). If properly constructed, your vision and roadmap will likely elicit the funding and resources needed to meet your objectives. At the same time, a thoroughly developed roadmap may also be rejected due to funding, timing or resource constraints. Ideally though, your roadmap will at a minimum reflect consideration of these obstacles with potential alternatives to reach mutual ground.

If successful in following the mariner's path to this point, you've secured funding for your endeavor. However, before the leader goes any further, they must understand and meet the requirements of servitude to their benefactor. In other words, what is the risk appetite of your manager, senior management, the board of directors, the security oversight committee, shareholders and clients? What is your own risk appetite and that of the employees and contractors working with you to execute the roadmap? What are the expectations of all parties regarding the execution of the roadmap (e.g., timeframes, durations, etc.)? Unless a common goal or purpose, along with reasonable expectations, is shared among all parties, the leader may be forced to make distracting course corrections throughout the execution of their roadmap with the potential net effect of never reaching their intermediary destination. Experienced leaders and mariners alike are aware of the pitfalls associated with juggling the appetites and opinions of the many parties involved in the execution of a vision. As such, they brace for the inevitable whispers of discontent and/or open displays of mutiny; especially when the trip fails to meet the stated or unstated expectations of one or more parties.

One sure fire way that ancient mariners sought to avoid conflict, at least immediately among the crew members, was to ensure adequate supplies for the journey based on an expectation of expeditious arrival while balancing the expenses associated with oversupplying, selecting the wrong supplies or purchasing costly supplies that could have been acquired at a more competitive rate. If expenses and supplies are managed properly, there will be little waste left at the end of the journey and a positive return on investment may be demonstrated. However, if poorly planned, a mariner may cut a trip short to request additional investment and resources from the benefactor. If the latter occurs frequently, members of the venture may begin questioning the mariner's ability to plan.

Security leaders must factor supplies, expenses and return on investment into their roadmaps. They must know when they intend to complete each leg of their journey (e.g., what calendar quarter, what fiscal year, etc.), how many and what types of resources may be required, what expenses (e.g., capital expenses, professional services, training, etc.) are to be expected, and what contingencies or dependencies exist that may lengthen or shorten a deployment.

Leaders should also tie these roadmap items to the various drivers communicated to the investors and stakeholders, as reasons for investment in the venture. Examples drivers may include regulatory requirements (i.e., HIPAA/HITECH, PCI DSS, Sarbanes-Oxley, State Breach Disclosure, etc.), shareholder profit expectations, client or societal changes in the protection of information, as well as advancement in new technologies. Even if you lack a method to precisely forecast dollar amounts, use ranges or approximations to provide program investors with a general sense of cost. For example, what is the potential duration of a project: Low (< 300 hours), Medium (<1000 hours) or High (>1000 hours)? You could do the same for expenses: Low (<$100k), Medium (<$250K), High (>$250k). Other categories for consideration might include: Product Differentiation, Regulatory Requirement, Resources Availability, Reuse, or Environmental Benefit.

The intent of including these items is to demonstrate consideration of the factors that are important to your business. It also ties the security roadmap with business plans, thus moving the security program from expense center to revenue generator. In addition to selecting meaningful categories for your business, also use approximations that are in line with ranges typically used at your company or as set forth by your finance department.

At this stage of the game, the mariner should be well under way navigating the open seas and executing their plan. They will have to correct for errors along the way as variables shift. They will routinely examine navigation tools such as the compass, celestial references and dead reckoning to determine location and progress. The use of dead reckoning—while still taught in some disciplines (e.g., ship captains, pilots, etc.)—is unfortunately lost on security practitioners.

Dead reckoning is useful throughout the execution of a roadmap as it provides perspective and a sense of accomplishment for leaders executing multiple deliverables. Dead reckoning enables a practitioner to find their present roadmap location by looking at the last milestone achieved and calculating average delivery intervals and general project direction. It helps to answer the questions of where am I (waypoint), what have I accomplished (consumption of resources), and will I meet my remaining goals (destination).

Whether using dead reckoning or alternate navigation instruments, the purpose is still the same. Mariners and leaders must routinely monitor instrumentation (metrics) and key performance indicators (measures), while maintaining progress reports (ship logs) so that overall progress can be demonstrated to benefactors. Failure to record and communicate activities over the course of a journey, including decisions to make course corrections, may place an individual in an indefensible position.

One such course correction that is often required on the open seas is due to weather complications. The skilled mariner routinely looks to the horizon for black clouds or changes in weather patterns. Failure to account for anomalies and make adjustments to avoid conflict may result in a ship wreck.

The same is true for leaders and therefore, they must continue to read, discuss and think. If a leader has become so rigid in the execution of their roadmap that they're unable to make course corrections when conflict is certain, then they will likely fail to meet their vision. A proper roadmap requires maintenance and routine revision when you're at the helm of a security program to again account for changes of any nature. For example, you may have selected a vendor that is now in the process of being acquired. Will their acquisition result in product changes or support expectations? Will you as the leader still execute on your vision when doing so may result in inadequate technology or continuous calls to the vendors support line? Do not be so rigid in your roadmap that you can't accommodate changes along the way.

Every step that a successful mariner takes will ultimately lead to an intermediary destination and a sense of fleeting accomplishment. A destination is intermediary in that a return trip will be required to demonstrate that an exotic destination was reached and the interests of the benefactor were upheld. As such, the mariner must now make the return trip thus establishing a new intermediary destination. He must incorporate the knowledge he has gained over the course of his journey, update maps, create research reports for examination upon return, gather samples and riches, plot a course, outfit his ship, and navigate and error correct along the way to the next intermediary destination and sense of fleeting accomplishment. For certainly, if both the mariner and leader are successful, they will be commissioned over-and-over again in support of all vested parties.

About Todd Colvin

Todd Colvin is the Director of Enterprise Data Security for Paychex, Inc, a Payroll, Human Resource and Employee Benefits service provider. In this role, he has international responsibility for the protection of all corporate and client information assets.

Prior to joining Paychex, Inc, Mr. Colvin was the Homeland Security Manager for Sprint and served on numerous Washington, D.C.-based committees-including the President's National Security Telecommunications Advisory Committee (NSTAC), the Network Security Information Exchange and was the resident representative to the National Coordinating Center for Telecommunications (NCC) and the Telecom ISAC. In this role, Mr. Colvin was responsible for the coordination of communications restoration during the hostile 2004 Atlantic hurricane season. When not responding to National Security or Emergency Preparedness (NS/EP) events, Mr. Colvin participated in several task forces including the TRUSTED ACCESS TASK FORCE to address Screening, Credentialing, and Perimeter Access Controls. Additionally, he developed and delivered a report addressing communications preparedness for National Special Security Events (NSSE).

Mr. Colvin is a dedicated security professional that holds many certifications including the CPP, CISSP, CISA, CISM and GSNA. Mr. Colvin holds a bachelors degree in Information Security and Assurance. He is also a 2007 graduate of the ASIS Security Executive Program at the Wharton School of Business and an Eagle Scout.

SECURE

Secure by Design: A Foundation to Software Innovation

Al Zollar

General Manager – IBM Tivoli
Headquarters: Armonk, NY
Security Leaders Summit Northeast 2010 – Platinum Sponsor

Through world-class solutions that address risk across the enterprise, IBM helps organizations build a strong security posture that helps reduce costs, improve service, and manage risk. IBM solutions are informed by the IBM X-Force research and development team, which studies and monitors the latest threat trends including vulnerabilities, exploits and active attacks, viruses and other malware, spam, phishing, and malicious web content. In addition to advising customers and the general public on how to respond to emerging and critical threats, the X-Force also delivers security content to protect IBM customers from these threats.

....

I am sure that I do not have to explain to you how your lives are getting more complex. The pace of change in business, politics and technology is accelerating exponentially. We are not talking about mild, episodic change. We are talking about "turbulent change." Economic disruptions, cyber attacks, political upheaval, technology leapfrogs, and natural disasters can occur almost without warning. Even anticipated change, driven by new partnerships, mergers and acquisitions, or management initiatives, requires leaders to make decisions in the face of uncertainty in order to ensure continuing economic growth and expanded stakeholder value.

In the face of this uncertainty, how do we anticipate and prepare for everything that might happen? Organizations have to learn to embrace change, and thrive on it. Every person knows all too well the complexities and challenges that technology is creating in a world that is becoming more aware, more interconnected and smarter every day.

Over the last two years, you've heard IBM speak about a concept we call "Smarter Planet". Some people refer to it as our "Intelligent World" while others refer to it as the "Internet of Things", but the concept is the same. Our planet grows smarter because of the sheer magnitude of the data that we can collect about the events and activities in our everyday lives, our ability to interconnect, collect, and share that data in a world where billions of devices have built-in intelligence, and because of our ability to use this data to understand and control what is happening in the environments around us. This year, 2010, marks the beginning of a new era. In these next 10 years, more than ever before, we will find ways to make the collection and analysis of this data actionable and change the way people live for the better. These changes will be driven by the development of leading edge software, because on our smarter planet, software is already changing the way people live. All across our planet, businesses, governments and industry are using software in new ways. Today, more than ever, organizations use software to enable every facet of their business. With advanced software and analytic tools, they have the capability to extract value from data- to see the patterns, correlations and outliers. They can uncover hidden patterns in data to analyze, predict and control events. Problems can be recognized and remediated before it's too late.

But with these new models and ways of working come both old and new challenges. For decades, companies have faced cultural, organizational and technical barriers that have limited their ability to address critical business processes that impede the quality and speed of software delivery. They lacked the expertise and capabilities for product and service innovation through the integrated processes of design, delivery and management of software engineered into intelligent devices and services. The answer to this challenge lies in the linkage of software development processes to service delivery, support and operational processes. As services become an integral part of business and system design, more comprehensive management of the service life cycle becomes a key requirement. In fact, in 2009, the Aberdeen Group published a study that showed that Best-in-class product companies are those that build a strong competency in software delivery. These Best-in-class companies achieved a 25% decrease in product development time, are 19% more likely to meet revenue targets than the industry average, and achieved 4 times more embedded software than their competitors, with 50% fewer defects.

Simply put, Service Lifecycle Management means that there is a business process that governs from the initial stages of project planning, right through to the billing of the users or customers. From a security perspective, it would ensure governance and risk management throughout the stages of Requirements, Planning and Analysis, Development, Testing and Operations. It requires that the software development and operations teams share a common automated lifecycle process and tools, an integrated platform of development and operational tools and a federated database of information for a shared context in visualizing, measuring and transforming performance. The business value of this process is not just improving the time to develop, test & deploy new applications. It can reduce operational risk and improve audit posture, reduce cost, improve efficiency and accuracy, simplify deployment and reduce duplication of effort and data. Beyond all this, it can be game changing for the companies where service lifecycle management has become a key element of product differentiation and end-user or customer experience.

An early historical example of a game changing impact of closing the gap between design, development and operations is the Japanese auto industry from 1960 to 1980. By applying lifecycle principles and introducing statistical control and scheduling techniques, firms like Nissan and Toyota radically improved the efficiency and throughput of auto design, manufacturing and delivery processes. Design, operations and customer service were linked early in the lifecycle process so that product development was refined to ensure the teams aren't building the wrong things. Planning and governance processes were simplified to support quality, responsiveness and adaptability early in the development process and resulted in production at speeds previously thought impossible. Divisions moved away from a top-down, hierarchical design, increasing collaboration between design and operations and introducing a stronger focus on customer feedback. We all know the results. In 1960 US car manufacturers produced and sold 7.8 million cars worldwide, while Japanese auto makers produced 480,000. In 1980, US manufacturers were stagnant at 8 million cars sold, because the Japanese companies had seized the growth in the market and produced and sold 11 million new cars that same year. They had been selling millions of cars into the American market, crippling the economic growth of the Big 3 car makers during the decade of the 1970's and contributing to a major recession here in the US. And they produced all of these cars only requiring 1 employee for every 50 cars manufactured, while the US companies required 5 employees for the same production.

More recently, here at IBM, we worked with a major European Petroleum Company that was dealing with long service downtimes caused by problem diagnosis issues. Inefficient access to resources and information was impacting time to restoration and causing poor client satisfaction and negative revenue growth. We helped them implement a common service lifecycle process, that allowed them to integrate their service design with client requirements, automate change and configuration management, and have simplified asset synchronization and traceability between development and operations. The result was significantly reduced downtime, faster root cause analysis, happier customers and revenue growth.

Similarly, we worked with a world-wide financial services conglomerate, whose IT operations team was drowning in a constant stream of change requests across services designed by 25 different application teams. Their operations staff resource budget was frozen and they could not keep up with all of the issues reported about performance issues, deployment errors and manual workarounds and had a current backlog of 2,500 priority one requests. We helped them implement a service lifecycle process that's discovers the interdependency of assets in their live topology so they could create a reference architecture. They also automated their workflow so that it could be re-used time and again for a repeatable and 100% accurate deployment. The result was a vastly reduced backlog of issues, cost savings through consistent deployment, compliance and traceability of work activity and the ability to capture, document and formalize best practices.

These are examples of success stories where companies have overcome the challenge of business complexity that was driven by more and more software innovation. The smarter planet is complex and we must ensure that our organizations are pro-actively identifying all the different types of challenges and risks to our business, and that we put plans in place to address any threat or disruption. Complexity is an issue that here to stay. In the area of IT security, complexity creates shadowy places where people with bad intentions can do a lot of damage. Complexity creates vulnerabilities. We cannot have a Smarter Planet without understanding the risks associated with all of our new capabilities and the new services that we are going to build. What has changed about security risk management in this new, smarter world? Well, human beings are becoming more reliant on security in our everyday lives and security is now seen as a right of every individual.

This is a simple concept, whether you are dealing with cyber security, protecting a critical infrastructure, developing new software applications or dealing with the issue of privacy and identity. In all cases, we always have to be ahead of the bad guys. We have to design security into the services that will drive a smarter planet. That is the essence of what we are trying to do, because the point in history at which we stand is still full of promise and danger.

To realize this promise of future innovation we will depend upon the safe and dependable operation of the systems that will gather, transmit, and analyze data, communicate and act upon the results.

This type of security, this type of safety, is not something that can simply be bolted onto the new services of the smarter planet as an afterthought. It must be considered from the first requirements to the final implementation, and it must be inherent in the capabilities that are brought to bear as complex problems are solved. The reliability of these solutions must be Secure by Design.

This concept we call "Secure by Design" means that cost-effective security begins with the creation of secure systems from the start. Time-to-market, maintenance, and the devastating costs of publicized breaches are reduced through the benefits of integrating secure practices early in the development lifecycle. It is a long-standing axiom that functional defects identified during system development are orders of magnitude less costly to repair than those found in production systems, and these benefits and savings are even higher when it comes to security. Current models show us that the average data breach costs an organization roughly $6.6M, and that the average cost per lost customer data record is over $200. These numbers are staggering. Security vulnerabilities within some smarter planet systems will be even more destructive if those systems manage critical infrastructure, and failure can disable entire regions or worse, jeopardize lives.

THERE ARE FOUR ELEMENTS TO SECURE BY DESIGN:

1. The first is an understanding of the design point for security, that is, if we were to declare a service or application to be "Secure by Design" what attributes does it possess-what is the definition by which security and dependability would be assessed for that component or system?

2. The second element is knowledge of the threats in the solution's operating environment, and of the likely risks that a system or component will face once deployed into the market. In order to be functional, cost-effective, and usable, designs must incorporate a rational view of risk that balances the

urgency of security. The dangers are constantly evolving, and maintaining a current view of their relative impact, likelihood, and mitigation, requires a deep level of commitment and knowledge. For example, IBM's X-Force has for many years, specialized in researching current and emerging security threats. They produce reports that are cited widely in journals and presentations on the topic of the evolving threat landscape. Addressing the relevant and pressing risks for a particular system or component begins with this understanding that we assist our clients to obtain.

3. The third security element is structural. Part of a secure design is the utilization of enabling technologies that provide or enforce secure behaviors. Access management, log-in, encryption, and authorization are all examples of design components that must be considered and appropriately contained in modules that will be internetworked, which will handle sensitive data, or which may ultimately control actual operation of systems.

4. The fourth element in this process is the ongoing validation of the secure design. System development history has shown us that there is a natural tendency for implementations to veer from their original designs, and that constant reinforcement of design objectives through testing and assessment are the most practical means of arriving at a deliverable with the proper attributes. Security is no different, but can be more difficult to assess. Security, particularly at the coding and implementation level, is not a widely understood or practiced discipline. Its inclusion in the set of critical deliverables will only be possible as organizations simplify and automate security checking in ways similar to those employed for functional and performance testing.

Each of these four elements is important. Without context, an approach to secure by design can be insufficient, or overkill, or become out-of-date very quickly. Without knowledge of the use of enablers that empower a secure application, important areas can be left inadvertently exposed or unprotected. Without assessment, it is almost impossible that the security that was intended will ever emerge from the other end of the cycle. The four support each other, and together support the security of the system.

Securing these components and systems from their inception produces a flexibility and sense of assurance that fuels the growth and adaptability of the Smarter Planet. Early implementations of smarter projects are only the beginning of the potential for integrating information and technology to solve fundamental infrastructural problems. Systems which today are gathering information to optimize a Smarter City may well be repurposed in the future to bring smarter health care or smarter communications to the same area. By designing the core components with security in mind, leveraging them into a new area of use becomes much more straightforward, eliminating the need to re-engineer the component for the next role it may fulfill.

Organizations which have already adopted a secure by design methodology have seen very positive results. One major telecommunications firm has gone so far as to apply the knowledge of their relevant threats and the operational implementation goals of their software components to devise an automated testing regimen that is kicked off regularly with the software build. The information generated has already been tailored by the security team, and the results are regularly reviewed to ensure relevance and continuing accuracy. In the interim, each build automatically assesses the security of the software, and forwards any newly found vulnerabilities to the appropriate development groups for remediation. This integrated process has led to much faster cycle times, decreased rework, and a far better performance during rigorous pre-deployment certification.

"Secure by Design" as I've described it, relates to assembling the knowledge, tools and processes to generate components and systems that will perform reliably and securely, through efforts at all phases of the construction lifecycle. Equally as important is having a service operations and security operations perspective that is secure by its design. You have to be able to see what's happening with the infrastructure and be in control. In a controlled environment, we strive to have everything in a desired state. You have to be able to visualize what is happening right now in that desired state and you have to make sure that you stay within the control limits and understand the risks that are associated with moving in and out of that desired state. Control also requires automation, because we've seen in our work with clients worldwide that vulnerability surfaces from people choosing not to follow repeatable defined processes. This is a very powerful value proposition and it resonates with businesses, industries and governments across our smarter planet; it's called Service Management.

At our Pulse conference this past February, we deeply examined this notion of Integrated Service Management. This idea resonates with our clients because it's a real value proposition that starts with a core premise, that the most important thing our clients do with technology infrastructures they deploy is deliver a service that supports a business process, that is important to a business outcome. And what's really important about that service from a security perspective is that it's built on a set of physical assets with built-in intelligence, storing, processing or collecting huge amounts of information, that are network connected. It is important to note that these intelligent and connected assets span multiple classes. They can be transportation vehicles, they can be traffic control devices, they can be buildings and, of course, they can be digital assets, like design documents or software that supports composite applications; all of these things can now be part of the smarter services you deliver. A key element of smarter security will be the capability to manage all of the assets across this spectrum with one set of software capabilities, because you have to be able to see what's happening with the entire infrastructure from both a service operations and a security operations perspective.

Integrated Service Management also includes the integration of the lifecycle management processes I described earlier and best practices from different sources, like ITIL, the IT Infrastructure Library or COBIT and ISO 27001, from a security control stand point. The integration of these best practices allow organizations to make sure that they are functioning properly within a set control limit, to deliver the services that are expected or mandated, in an increasingly interconnected and complex world.

So, at IBM we talk to our clients about this idea of Integrated Service Management as the "operating system" for the Smarter Planet. Interestingly, while it has been found that organizations today will turn to any number of best practices for guidance on which controls to implement – often driven by regulations, industry, geography, or corporate culture - it is adherence to the disciplines of Integrated Service Management, that we see setting the highest performers in the area of security management apart. Integrated Service Management drives down the cost of implementing and maintaining service infrastructure, it minimizes human error, and it provides the data and insights to measure and prove compliance. It is also integral to the creation of software development processes, resulting in the security I referred to earlier that is built into the software applications we will use to create new services for the smarter planet. Every business and every government, regardless of their industry or home continent, will need strong integrated service management capabilities to deliver the Smarter Bridges, Smarter Utility Grids, Smarter Farms, and Smarter Banks of our future.

In fact, it seems clear that the most successful enterprises in the next decade will embrace Secure by Design, and Lifecycle Management with Integrated Service Management as the foundation of their business processes. And those that do so will be able to seize new opportunities to innovate securely and cost effectively. However, you can't move to a paradigm of Smarter Security by yourself. It's too big a job. You need partners who can help you get there-partners who can deliver the technology, business expertise and global service capabilities to achieve Smarter Security. At this point in history we can realize the promise of tomorrow and minimize the danger by achieving strong security, which leads to strong prosperity.

About Al Zollar

Al Zollar, general manager of IBM Tivoli Software, is responsible for the strategic direction and ongoing operations for the Tivoli brand, which manages today's dynamic infrastructures, giving customers the ability to manage resources and risks, optimize human capital and manage service levels and business processes.

Since joining IBM in 1977 as a systems engineer trainee, Al has held several high-level positions, including serving as general manager for eServer iSeries, Lotus and IBM's Network Computing Software Division.

About The Editor - Mary Lou Heastings

As CEO of Executive Alliance, Mary Lou Heastings has responsibility for guiding the continued expansion of the company's portfolio of technology related executive summits, awards, roundtables, and custom programs across the country. Ms. Heastings launched the publishing arm of Executive Alliance to promote the successes of executives in different industries. Ms. Heastings has significant experience in management, information technology, finance and operations.

About Executive Alliance

Executive Alliance is the premier provider of a national network of executive technology programs that recognize the achievements of leaders in different industries. These programs facilitate the building of deep relationships within a network of peers, provide visibility for executives and their companies, and offer access and insight into the people leading these industries.

Executive Alliance has produced executive technology summits and programs across the U.S. in cities including Atlanta, Boston, Chicago, Dallas, Houston, Las Vegas, Los Angeles, New York, Philadelphia, San Francisco, Washington DC., and internationally, London and Bangkok.

Visit **execalliance.com** to learn more.

About The Designer

Scott Cathey is a Graphic Designer and Illustrator based in Atlanta, GA. As owner of Scott Cathey Design, he specializes in brand identity and logo development, print collateral, web design and illustration.

With over 20 years of experience and a passion for solving creative conundrums (large and small), Scott helps his clients communicate effectively to achieve success in today's marketplace.

Examples of his work can be viewed at **scottcathey.com**

ORDER MORE COPIES:

execalliance.com
amazon.com
or call 678.445.1919

Order additional copies of "SECURE" for yourself, your team, or peers. This is a perfect gift for a security professional.

Interested in learning more about our Security Leaders Summit[SM]?

itsecurityleaders.com